I am Spring. After the cold of Winter, I'm the season when the Sun starts to slowly warm the land, and everything begins to grow again. Flowers open and show their bright colors, animals and insects come out of hiding, and new life is born into the world.

One way that people can tell it's the first day of Spring is the light of day lasts as long as the dark of night. This is called the Spring Equinox.

During Spring, each day the Sun rises a little earlier and sets a little later in the afternoon, making the days longer and brighter.

Brighter weather means the dirt is warmer, so seeds sprout and push their way up toward the Sun.

Spring is one of the growing seasons when seeds grow into vegetables that can be picked and eaten.

High up in the sky, when warm Spring air mixes with cold ocean air, it can cause bursts of rain showers, watering the land and helping things to grow.

Bees and other insects come out of hiding and visit flowers for a sip of sweet nectar and a bite of pollen.

Many birds that flew away to avoid the cold Winter return for the warm sunshine of Spring skies. When birds travel to far off places for better weather and more food, scientists call this migration.

As Spring warms the land, the sound of birdsong fills the air. Birds chirp, tweet, and trill, calling out to meet one another, and to claim the areas they want to live.

Spring is also the time for building nests and laying eggs!

Spring sunshine and the smell of growing foods wake up sleepy animals that rest during the cold months.

Bare branches on trees begin to sprout new leaves, and some will also grow blossoms. Blossom is another way of saying flower.

A gentle Spring breeze moving through blossoming branches can make it feel like it's raining flowers.

Plants, like trees, are very helpful with keeping the air clean. They remove pollution through their leaves and catch dust and other floating particles on their branches and bark. With all the new growth in Spring, there are lots of plants to clean the air.

One way that people help to keep the air clean is by planting new trees and taking care of the plants that are already here.

When the day becomes longer than the night, and the leaves begin to sprout on the bare branches, and bugs and animals that were hiding start to reappear, get ready, because that means Spring is here!

What's another name for the first day of Spring when the day is as long as the night?

What happens when warm Spring air mixes with cold ocean air?

What are birds doing when they fly far away for better weather and food?

What is another word for flower?

I Am Spring
Copyright © 2019 by Rebecca and James McDonald

All rights reserved. No part of this publication may be reproduced, stored, or distributed in any form or by any means, electronic or mechanical, including photocopy, recording, or any information storage and retrieval system, without prior permission in writing from the publisher and copyright owner.
ISBN: 978-1-950553-17-4
First House of Lore paperback edition, 2019
Visit us at www.HouseOfLore.net

Made in the USA
Columbia, SC
08 March 2024